Rock Hounding for Beginners

Contents

Introduction

Do rocks, minerals, crystals, and gemstones fascinate you?

Would you want to collect different types of rocks, gemstones, and minerals but are not sure how best to do it?

Would you want a detailed, step-by-step guide on rock hounding, which entails collecting different types of minerals, rocks, gemstones, and other items underneath the ground?

If yes, then keep reading.

Exploring the world around us is part of our humanity and something we cannot shake away. Rockhounding, which involves exploring the earth and everything buried beneath it, takes hobby exploration to a new level.

Rockhounding goes beneath the surface and looks at what lurks underneath the earth, unveiling the earth's long history the deeper you dig.

So, if you are interested in unraveling the earth's history through rocks, you might be wondering:

- How do I get started in rockhounding?
- How do I identify areas best for rockhounding?
- What minerals and rocks will I find in rockhounding?

If you have these and many more questions about rock hounding, this book will answer them and provide you with valuable information you can use as you get started in rock hounding.

In this book, you will learn about the following:

- What rock hounding is and what it entails
- How to prepare yourself for rock hounding
- Tools you need for rock hounding
- Best minerals, rocks, and gems to look out for

And so much more.

So, let us begin!

Chapter 1: What is Rock Hounding and How Do You Get Started

Let us start by understanding what rock hounding really is and what it entails: Rockhounding is the act of exploring the environment in search of rocks, minerals, gemstones, or crystals. It is also called amateur geology and is a recreational activity where people who are not professional geologists engage in the study and search of the mentioned items in their natural environment.

People who engage in rock hounding do it for various reasons.

For example, some rock hounds are jewelry makers seeking stones for some of their pieces. Some people might participate in rock hounding to understand their local history and geology. Others could simply enjoy rock hounding as an added challenge to their hiking trips.

Whatever reason you have for engaging in rock hounding, it can be a very rewarding and fascinating hobby.

How to Get Started

As with any other field, getting started in rockhounding involves extensive research. First, you need to understand what is available in your area. This means knowing what rocks and minerals are available around you and how far you will need to travel to collect whatever you need.

The quickest way to know the kinds of rocks available in your area is by doing an internet search. You can simply type in 'rock hounding in (insert your area).' This should yield results for you. In the best-case scenario, you will get some locations and maps from people who have been to that place before.

You can also visit your local library or bookstore to get some reading material. If you can find books that detail rocks in your area, that would be great. These books will provide an excellent reference for the kinds of minerals, crystals, gems, or fossils you will find in your area. These books can also provide information on areas considered safe for rock collecting.

An even better way to do your research is by joining a rockhounding club near you; after all, there is no better way to learn about rock hounding than from other, more experienced rock hounds. The mineral collectors club or rock hound clubs found locally will teach you more about your locale and the best approaches to collecting the said rocks.

When you can benefit from the knowledge of people who have been there and done that, your first rockhounding trips can be much more productive and less strenuous than if you try and go it alone. You can find a list of local mineral collecting clubs on the American Mineral Federation list.

Other valuable sources of rockhounding information you can look for include:

State Natural Resource Agencies

Most states have agencies that are responsible for geology and mining. Often, these agencies will have information about the local rocks, gems, minerals, and fossils and will often willingly provide information if requested.

Museums

Geology, natural history museums, and other such areas are suitable if you want to study specimens of rocks you can find in your area. They are a great place to get a close-up view of the kinds of rocks you will be looking for as you venture out. You will also get education programs to prepare you for your expedition.

Rock shops

Rock shops specialize in rocks, minerals, and gems, which can make them a great place to see high-quality rocks you may have previously only seen in images and meet other people who are enthusiastic about rock hounding as you are.
The staff of rock shops can also be a valuable source of information on local rockhounding areas and local rock clubs.

Local university

If your local university has relevant programs for rockhounding (geology, paleontology, or mineral engineering), you can find resources that will prove invaluable to your rockhounding desires.

Rock, gems, and mineral shows

Yes, these exist. Most of these shows will usually have a low entrance fee and give you the best chance of seeing a variety of rocks, minerals, and gem specimens.

Safe Lands for Rock Hounding

There are three major areas where you can dig for rocks without getting into trouble. The first is public land, where you will not need permission or permit to dig the land. You should gather information on this during your research to ensure you have some direction, know where you are going and that the area is safe for digging.

The second is pay-to-dig sites. These sites are mines that allow people to dig through the mine for a small fee. Many of these sites charge amateur rock hounds a reasonable fee, usually ranging from $5 to $50 per day. Additionally, many of these mines are family-friendly; thus, you can take your children there and get them started early on this rewarding hobby.

The third, more complicated place to rockhound is private land. On private land, you should not dig on the land before getting explicit permission from the owner. Digging on private property without permission from the owner might lead to legal trouble, including but not limited to prosecution for trespassing, or worse, you might get hurt if you start digging on land belonging to a trigger-happy Joe sixpack.

Aside from that, you should also not dig someone else's mine claim. Mining claims will have markings, often corner posts or signage, indicating the claim. Thus, digging through such a mining claim without permission from the person who has laid a claim to the area is never okay. Besides being illegal, it also goes against the unwritten rule in rock hounding, which frowns upon digging someone else's work.

Also, note that public land such as State and National Parks and National Monuments are out of bounds for digging and collection, often because of their legal protection.

Rules and Regulations

Rockhounding has rules and regulations, many of which are the purview of the Bureau of Land Management (BLM).

The following are rules you should be aware of before you go out rock hounding:

- The land from which you collect rocks, minerals, or gems should be under BLM management.
- In rock hounding, you cannot use motorized, mechanized or heavy equipment or explosives. You can only use hand tools such as shovels, picks, or hammers. You are also free to use metal detectors.
- Rockhounding on recreational sites is prohibited unless the site is a designated rockhounding area.
- If you decide to pull resources with other rock hounds, you should not collect a mineral, gem, or rock piece larger than 250 pounds. To obtain a piece larger than 250 pounds, you will need to be in contact with the local BLM offices.
- When rock hounding in Wilderness Areas, you must restrict your rockhounding activity to surface collection.
- Historical artifacts, for example, items belonging to Native Americans, such as arrowheads, parts of pottery, human remains, and such, cannot be collected. Additionally, you cannot collect vertebrate fossils. So, collecting remains from creatures with a skeletal structure, like mammals, fish, dinosaurs, or other creatures with bones, is prohibited. You can only collect rocks, mildly-precious gems, mineral specimens, some invertebrate fossils and petrified wood.

The U.S National Forest Service also allows for rock hounding in American forests, but they also have some rules and regulations:

- When rock hounding in the U.S forests, you are allowed to collect small amounts but a wide variety of low-value, common minerals and stones. Quartz, agate, crystals, and obsidian are some of the rocks you can collect for noncommercial use.
- You are allowed to conduct hobby mining activities, including recreational gold panning or using metal detectors to find gold nuggets and other metals occurring naturally in the earth.
- The amount of specimen you can collect is limited to 10 pounds.
- You are not allowed to rock hound in all parts of the forest. Some lands within the National Forest are closed due to wilderness designation.
- Collection of vertebrae fossils is also prohibited. Furthermore, you cannot collect historical artifacts such as pottery fragments, arrowheads, or other things.
- Before going out for rock hounding in any U.S forest, ensure you contact the U.S Forest Service for up-to-date information.

Rockhounding in Rivers and Creeks also comes with some regulations.

- If you are rock hounding in a river or creek designated as an Essential Salmon Habitat, you can collect up to one cubic yard of rocks, gravel, or wood.
- If you are rock hounding in a stream designated as a State Scenic Waterway, you will need a special permit called A Scenic Waterway Removal-Fill Permit. This permit lets you collect materials on the waterway. You are allowed to collect about 50 cubic yards of material per year. However, the permit will cover only the specific waterway for which it was designated.

The Oregon Coast is a popular destination for many rock hounds; thus, you need to be aware of the regulations specific to that park:

- Parts of the beach might be part of the Oregon State Park; thus, it will follow the regulations above. However, there are some general rules for collecting material at the beach that the park manager will inform you of.
- You are allowed not more than one-gallon container for every person per day of items such as agates or shells, stones, and fossils. For a year, only three gallons per person of material is to be taken away from the Oregon coast.

Now that you know the rules and regulations, we can focus on the tools you require and that are legally allowed for amateur rock hounding.

Chapter 2: Rock Hounding Tools and Supplies

Rock hounding can be as simple as walking around a piece of land picking up rocks that fascinate you. However, the more involved you become with the hobby, you will find that your desire for minerals and rocks extends beyond simply those you can easily find on the surface. So, walking around looking for surface-level rocks is no longer fulfilling. Instead, there is a desire to travel long distances and use special tools to dig through the surface to find some treasure.

We can segment the tools you need for rock hounding into several categories.

- Tools for protection
- Navigation tools
- Tools for collecting specimens
- Tools for determining collectible specimen

Tools for Protection

Rockhounding will involve many risky encounters with rocks, debris, and other things that could injure you. Thus, having the proper protective clothing is a great idea. Here, you need the following:

- **Goggles –** You will be doing a lot of digging; as you know, digging throws around a lot of loose earth. A reliable pair of goggles can protect your eyes from dust, metal fragments, and chips.

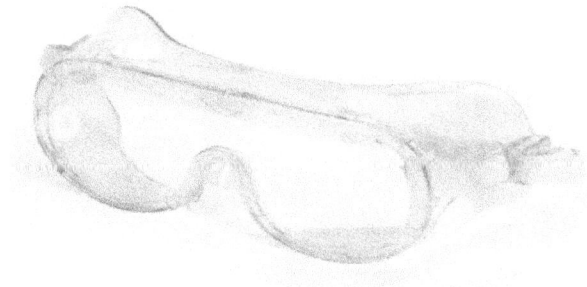

- **Gloves –** Gloves protect your hands from scraps and cuts from the metals and rocks in the wild when you are hammering away. They also keep your hands mostly clean.

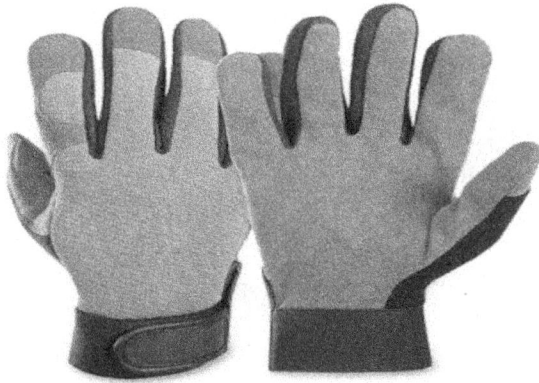

- **Hard hat –** While not often necessary in all rockhounding trips, a hard hat is necessary if your exploration takes you through areas like caves, cliffs, or any location that exposes you to the risk of falling rocks.

- **Proper footwear –** Boots with ankle support are great for rock hounding because they make it comfortable to move through the rugged terrain without risking twisting your ankle. Waterproof boots are a plus-to-have if you intend to tread areas with rivers, creeks, or lakes.

- **First aid kit** – Always have a first aid kit with items such as alcohol, bandages, and painkillers for treating minor injuries.

Navigation Tools

Your smartphone will probably have a compass and GPS maps you can use as a guide through your rockhounding trip. However, it is also wise to bring some traditional navigation equipment for moments when you may find yourself in an eventuality where cell service is unavailable in a certain area or location.

A handheld compass and a topographical map will come in handy; they can help you find your way to the rock collection sites and find your way back to civilization should you get lost.

Specimen Collection Tools

These tools include:

Rock hammer

A rock hammer or rock pick is a simple tool that you shall use to chip away or pry some small rocks. You can also use it on medium-sized rocks. Do not use a rock hammer on larger rocks.

There are different rock hammers, all of which will do different things on the field.

- **Estwing Big Blue** – This hammer is also called the mason's hammer. It has a square head and a chisel end on the other side. It is excellent for loosening hard soil and cracking rocks into small pieces, but it can also perform the function of a chisel when you find yourself in instances where you don't have one.

- **Chipping hammer** – Chipping hammers have one vertical chisel end and another pointed end. These hammers are used to make precise hits to small areas. For example, if you want to crack open a small piece of rock to access a gem inside it, a chipping hammer is the best tool here.

- **Geo rock pick** – Geo rock picks, also called paleo picks, are a hammer with a pointed end and another chiseled edge. Their most common use is to break apart sedimentary rocks or dig small holes in a cliffside. They are also great for digging small holes when kneeling.

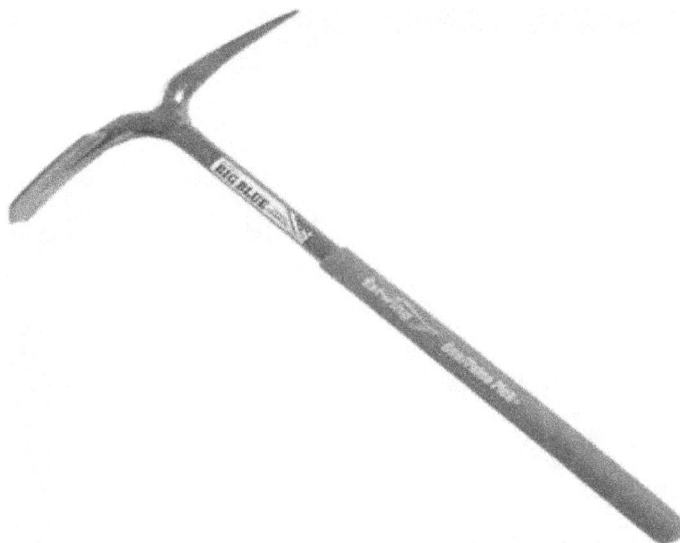

Geo/Paleo Rock Pick

- **Crack hammer** - A crack hammer is a small hammer that you can use to crack or break up rocks. You can also use it alongside the chisel to crack open rocks. It has two flat ends and looks a lot like a miniature sledgehammer.

Crack Hammer

- **Sledgehammer** – A sledgehammer is a large and heavy hammer that comes in handy when breaking up larger rocks into smaller pieces. So, a sledgehammer is the perfect tool for the job if you are rock hounding in an area with large, hard rocks that require a lot of force to break down.

Sledge Hammer

- **Cross peen rock hammer** – The cross peen hammers come with a flat hammer head and a chisel end. You can use them to break up your hand samples further and drive chisels.

Cross Peen Rock Hammer

NOTE: When selecting a hammer for your rockhounding adventure, ensure it has the following features:

- **Long handle** – A long handle provides increased distance from the area of impact to your hand, thus protecting your hand. However, hammers with long handles are more difficult to use than small handle hammers, making regular practice important.
- **Cushioned handle** – The cushioned handle provides better shock resistance from the vibrations when you hammer the rocks. This reduces the chances of your hands becoming numb after long use periods.

Chisel

Having chisels of different sizes and those that are both pointed and with wide blades is a great idea. The different sizes and edges will help you with different functions depending on what you want. For example, if you want to chip away only the edges of a specimen, then a wide blade, small chisel is the best tool here.

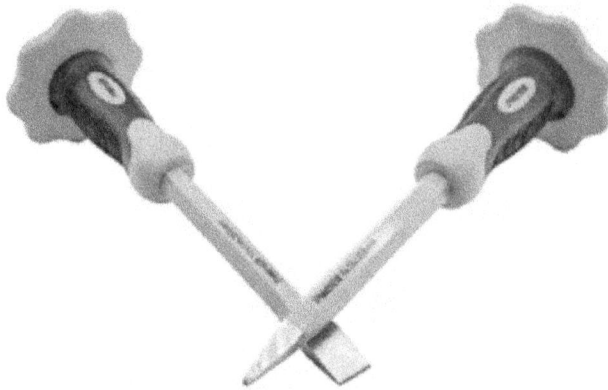

Crowbar

Also called the pry bar, this is a heavy iron bar with a bent end that you will use to move heavy rocks or debris that you cannot lift manageably. The pry bar should be long but not too long to make movement difficult. A crowbar of between 18-22 inches in length will be enough for the purpose.

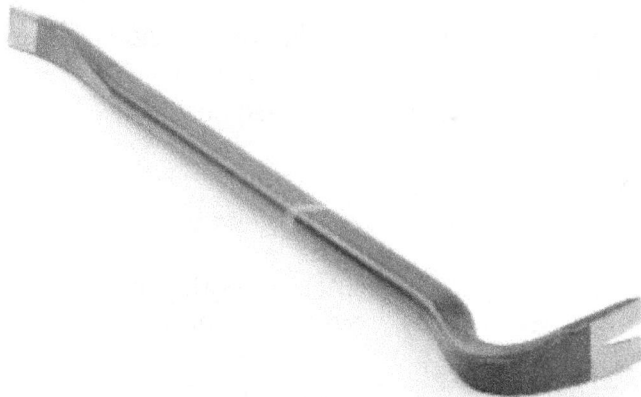

Cleaning tools

Tools such as a paintbrush, a small broom, and a spray bottle of water are crucial to rock hounding. These tools are useful in getting rid of grime and soil on a specimen for easier identification.

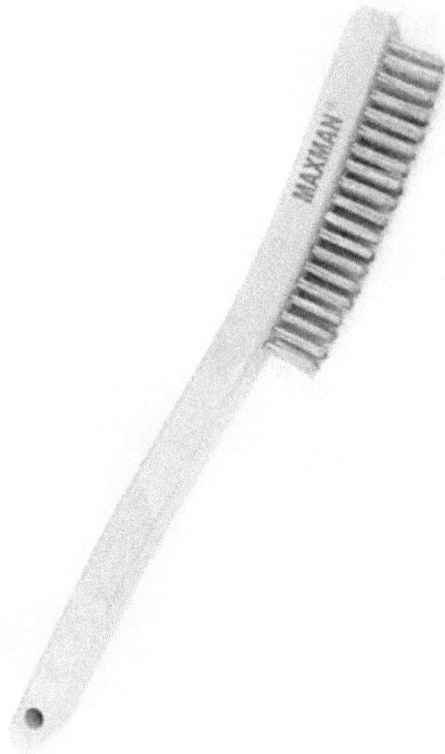

Hand tools

Various small hand tools such as a small pocket knife, colander, trowel, and smaller picks are useful for more precise work.

A colander

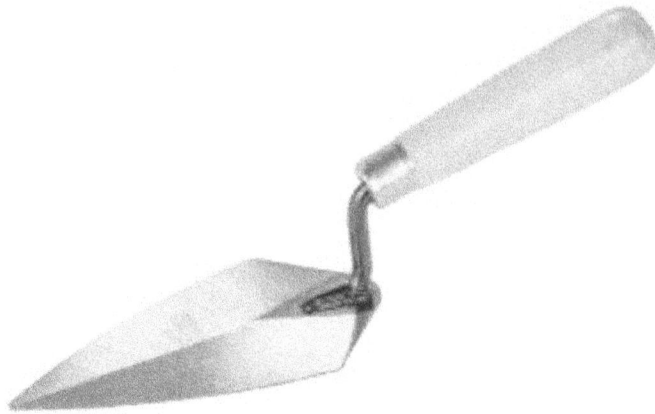

A trowel

Shovel

A full-size shovel is important in rock hounding as it is the tool you will use to remove dirt, grass, or rocks from the surface to reach your specimen. You can also use shovels to dig up holes in soft earth locations.

Carrying tools

You mustn't forget to take the tools you will use to store your collected specimen. A bucket where you can keep multiple specimens is a good option. Get durable buckets that can handle heavy loads and that have proper lids.

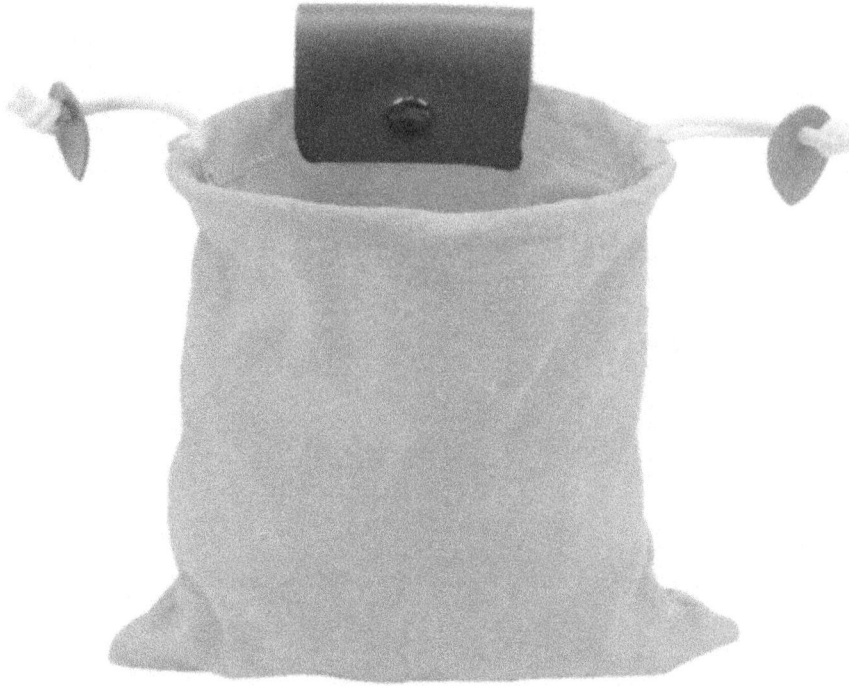

You can also use small boxes or tubes, especially when collecting smaller or delicate specimens. Prepare tubes or boxes of various sizes to fit an assortment of specimens.

Wrapping material is also useful, especially in keeping fragile specimens in mint condition. Usually, when putting the specimen in one area, they will grate and rub against each other, leading to them breaking or chipping. So, wrapping material such as newspaper or fabric is useful.

Tools for Determining Collectibles

Identifying rocks and minerals worth collecting can be daunting, especially with the naked eye. Therefore, you need to bring tools that can help you in the identification process.

- The first is your local field guide, a map that will guide you on where to go and what specimen you will look out for when in that location. This eliminates the need to go in blind.
- Magnifying glass – A magnifying glass is essential because it helps closely examine the qualities of the specimen you have found.
- Magnet – You can also use magnets to find iron-rich rocks such as hematite or magnetite.

Tools for Comfort

Aside from tools for the job, you should also prepare yourself for comfort in your rockhounding expedition.

Being outdoors will expose you to the elements; thus, you should do your best to ensure you do not succumb to the worst of the outside world. The following items are crucial (Remember that what you actually need depends on the prevailing weather, terrain, and duration of the trip).

- Rain gear
- Extra clothes and shoes
- Sun hat or bandana
- Food and water (quantity depends on the duration of the trip)
- Sunscreen
- Hand sanitizer
- Insect-repellent spray or cream
- Whistle
- Flashlight
- Toilet paper
- Trash bag

Chapter 3: Best Places to Collect Great Specimen

While we have looked at the lands where you are legally allowed to collect rocks, minerals, gems, and fossils, in this chapter, we will give you a detailed look into the best kinds of locations in these legally allowed areas where you can find what you are looking for.

While it might seem easier to just go into public land and begin digging, finding areas that are easily accessible, can be easily dug up, and have potential is crucial. But before detailing these locations, we need to look at what makes an area great for rock hounding.

Qualities of a Great Rock Hounding Location

You can and will most likely find semi-valuable or valuable rocks almost anywhere. However, the reality is that some spots are better for rockhounding than others, especially when you are new to the hobby. As a new entrant into the hobby, the last thing you want is to spend your first few hounding trips digging up nothing worth taking home; the hobby can get demotivating pretty fast when that happens one too many times.

Below then, are characteristics of a good rockhounding location to note when out in the field:

Exposed rock locations

A surface featuring many exposed rocks is a great location for a rock-collecting enthusiast. And this isn't limited to large outcrops; it also applies to smaller rock outcrops. As long as a location has rocks exposed on the surface, it has great potential. This is because these rocks will often be exposed due to soil erosion, which makes the surface a lot easier to dig up. Such areas also tend not to be overly grassy, thus making the terrain easier to navigate.

Different colored rocks

The earth forms from the surface to the rocks underneath due to several factors. Among these factors is acidic mineral solutions, which often affect rocks that form in a given area and depth. Due to these acidic mineral solutions, the rocks will often turn a shade lighter and continued deposits of different compounds from these solutions create rocks with different formations and build.

Thus, areas with different colored rocks might be a good place to find some gold nuggets or collect a sample of different rock and mineral types that you wouldn't otherwise have found at the back of your garden, for example.

Also, while at it, check out for rocks with quartz veins. Quartz veins are long, sheet-like crystallization of quartz material within a rock. You can identify quartz veins when a rock has a glassy luster running through it. Also, the rock breaks into curved shards rather than flat-edged fragments when cracked.

Fresh exposure

While finding exposed rocks is crucial, another sign of a great rockhounding location is that the rock exposure is fresh and the earth easy to dig through. A fresh rock exposure will often not have been subjected to the elements for too long.
So this means that the location should not have a big soil build-up or too much vegetation cover. And there are two reasons why these two factors are crucial.
First, fresh rock exposure means you will find higher quality rock or mineral samples that have not been degraded much with either rain, wind, or exposure to other elements, such as vegetation growing over them. Exposure to elements will lead to staining and discoloration or make a rock brittle in unnatural ways.
The other reason is that fresh rock exposures will not be immensely demanding. Freshly exposed rocks mean you will need much less digging to find the high-quality specimen you want. Many experienced rock hounds consider tampering with rock outcrops as an affront to nature, as you also contribute to making the rocks much more susceptible to the elements.

Safety

A rockhounding location is not ideal if it is unsafe to explore. Rockhounding is a hobby that will demand a lot from you, but you need to know when to make exceptions when your safety is in danger.
When you go out for rock hounding, consider areas that do not have a greater risk than usual. For example, it is okay to go to mountainous regions to look for quality rocks, but if the area has plunging edges and steep inclines, you shouldn't risk your life for some rocks, no matter how precious they may be.
Also, avoid locations where there is an increased risk of rock collapses. Besides steep mountain edges, these areas will also include long abandoned mines and shafts and areas with loose earth.
If you decide to explore caves, ensure you only explore caves that are large and wide enough for you to move through comfortably and that they are not too dark and isolated.
So, what locations often meet many of the above characteristics and are good for beginners?

Overturned soil

Areas with freshly overturned soil are a great place to begin your rockhounding journey, especially as a beginner. These could include construction sites or locations near newly-dug farmland (as long as you get permission from the farmers). However, note that many areas and locations with established farms have fewer rocks on or near the surface. Besides these locations providing a good place to begin your rock hounding, they will also not be too demanding in terms of equipment.

Quarries

If you want a great place to look for rocks, then a quarry is one of the best places for your purpose.

Indeed, quarries exist for the sole purpose of extracting minerals that many may find valuable. But first, before entering any quarry, ensure you have permission from the landowner to get in because many quarries in the U.S are privately owned. Many owners will often not grant permission to individuals due to liability problems. However, they might be more open to granting permission to groups so you could visit quarries with your rockhounding club.

Rockhounding in a quarry has several advantages.

First, a quarry has many dug and exposed rocks, which gives you access to rocks, minerals, and gems you would otherwise have struggled to reach using manual equipment. This means you can come across many rare rock types and layers to choose from as you go about rockhounding.

Secondly, since large machinery and vehicles need to access the quarry for different purposes, you can easily drive into and out of the quarry. Thus, accessibility should be a factor when considering the quarries for your rock hounding. However, this may not apply to long abandoned quarries.

River banks

River banks are great for rock hounding due to shifting water levels and running water that wears down the earth's surface, exposing new rocks. Thus, river banks give you a high chance of discovering new rocks as the water continues to wear down the exposed earth.

Sandbars or sandy banks make for great areas for rock hounding as they are usually on the inside of the river bends, which means the current flow will be gentler here, allowing for an accumulation of rocks and sand. This accumulation allows an assortment of rocks and deposits to gather.

The best way to get the most out of your visit to the rivers is to visit after a rainstorm after the water level recedes. The high volume of water pushed through the banks during a high storm will have brought new material.

Most of the rocks you will collect here will be smooth and rounded, thanks to the effect of water running over them for a long time.

However, ensure you keep safety in mind. Wear waterproof boots with great traction, and ensure you also bring someone else with you if you can.

A creek bed is a great alternative if it is impossible to find a river bank. Creek beds are the edges of creeks, and in creeks, the running water level will often be lower than what will typically be in a river.

This means that creek beds will be safer for you. However, because creeks are smaller in nature—at least compared to a river— many of the rocks you collect from here will be local because creeks do not have enough water or water flow to travel very far.

However, due to the lack of heavy water flow, you are likelier to find rocks of varying shapes and sizes and not just singular round rocks as you would in a river. Thus, rocks with crystallization will appear better preserved in the creek than in the river.

Road cuts

Road cuts are locations where large rocks have been cut through to give space for the passage of roads. This happens when building a road over the rock is not conceivable, either because it is too expensive or it might not make for a good road passage.

Road construction crews usually use excavators and explosives to make road cuts; thus, there is a high chance of collecting small rock samples of the sizes you want. You can collect the rocks on the side of the road, but if possible, also try to collect rocks from the top of the cut rock.

However, a major concern for road-cut rock collection is that there will often be very little room to park. Due to the nature of cutting rocks, the engineers will often only create room for the road, leaving little room for a shoulder to park on.

Thus, if there is no parking spot around the road cut, do not attempt to park there. Try to find a proper parking location away from the road cut.

Beaches

Beaches are to the sea and ocean what a river bank is to the river. They are a place where the water breaks, and the constant barrage of waves on the beach means you will find a lot of valuable rocks, minerals, or gems carried over from the ocean's depths. Aside from rocks and minerals, you can also collect sea shells.

The best time to visit the beach for rock collection is during low tides after a storm. Besides finding a lot of new and fresh specimens, the rainwater will have made the sand compacted, making it much easier to spot rocks and minerals without loose sand covering them.

Rock outcrops

Rock outcrops involve locations such as rocky hillsides, mountainsides, and cliffs. These locations differ from road cuts and quarries in that rock outcrops are naturally occurring, which makes it more likely to collect a wide variety of rock specimens.

Rock collection at the base of a rock outcrop can give you some great samples that give you a detailed idea of what the rocks above are made of. So, for example, collecting crystals or colored rocks is a sign that many of these are on the rock outcrops above them. Thus, there would be no need to break the rock outcrops.

Mine tailing

Tailings are materials left behind after the valuable component of a mined ore has been extracted. For example, if an ore containing iron has been dug up and the iron extracted, what remains is the tailing.

Mine tailing refers to the leftover material after an area has been mined for its valuable material. In mine tailings, mine companies often leave behind an extra rock and other debris from which you can find valuable material. This is true because mines, just like quarries, are only established in locations with something worthwhile. Thus, while what you may find will not be as valuable as the extracted ore, you are more likely to find something to make the trip worthwhile.

As with quarries, mine tailings are also easy to look through since someone else has already broken up the large rocks for you. Thus, you only need your magnifying glass and cleaning tools in hand to look for the best specimen in the location.

Chapter 4: Best Rock Hounding Sites in the U.S

Every state in the U.S has minerals, rocks, gemstones, crystals, and fossils. You can find some of these specimens in multiple states, but you can only find some of them only in specific states. Furthermore, in some of these states, you will need permission to collect the rocks and minerals.

Below are some of the most common and best rockhounding sites across the U.S

Crater of Diamonds State Park

Located in Murfreesboro, Arkansas, this state park permits you to dig up rocks and gemstones that you can keep. The park is 120 miles from Little Rock and charges a small entrance fee to access the park.

Located on top of an eroded volcano crater, Crater of Diamonds features over 75,000 diamond specimens (with colors ranging from yellow, white, and brown) that you can dig up and keep.

Besides the diamonds, you may also get some good gemstone specimens. Indeed, the visitors center displays some of the specimens you can expect to find in the location. The park also has interpretive programs for beginners, telling you what you can expect to find and why finding them is valuable.

Aside from rock hounding, you also get walking trails, campsites, and picnic sites at the Crater of Diamonds State Park.

Emerald Hollow Mine

Located in Hiddenite, North Carolina, the Emerald Hollow is distinctly known for being the only emerald mine in the world that is publicly accessible. The mine is at the foothills of Brushy Mountains, about 87 miles west of Greensboro. The mine is open seven days a week (from 8:30 AM-8:00 PM) but closes during bad weather conditions such as heavy storms.

The mine has naturally occurring gemstones, emeralds, topaz, amethysts, and other gemstones you can hound and take home. The staff is on hand to provide answers to your questions. The mine also has a shop where you can cut and set the gemstones you have found.

Gem Mountain Sapphire Mine

Located in Philipsburg, Montana, Gem Mountain is among the best places in the U.S to hunt for gems. The area is friendly to seniors and children because it is safe, and you do not need much labor to hunt for valuable rocks and gems.

The location is home to several rocks and gemstones, though the sapphire specimen is the most common gemstone you are likely to find at this location. However, the mine is open only four days a week, and you will need to call ahead to ask about reservations.

Jade Cove Trail

Unlike the other three locations above, the Jade Cove Trail is not a specific location for rock and gem hounding; instead, it is more of a hiking trail.

Also, unlike the specific rockhounding area, the Jade Cove Trail prohibits where to collect rocks and what kinds of rocks you can take away from the area. For example, you can hunt for rocks only during low tides, when stones wash up to the coastline from the ocean floor.

However, note that the location has a very rugged terrain; thus, shoes with great traction and that provide proper ankle support are a must-have.

Rainbow Ridge Opal Mine

Located in Virgin Valley, Nevada, the Rainbow Ridge Opal Mine is one of the costlier rockhounding locations to access, with an entry fee set at around $100. The location is also more suited to experienced rock hounds because finding opals, the most valuable gemstones you are likely to find at this location, is pretty hard.

Indeed, the opals found here are some of the most valuable gems of their kind and come in various colors, from colorless to emerald green to black.

The mine is open from May to mid-September, and you will need to arm yourself with gas and other personal equipment as such services are few within the expansive mine itself.

Gold Prospecting Adventures in Jamestown, California

Panning for gold nuggets is a must-do activity for any rock hound and the Gold Prospecting Adventures in Jamestown, California, is perfect for gold panning.

Gold Prospecting Adventures is located to the east of San Francisco (a 3-hour drive) and is a great place for a beginner due to several factors.

The first notable thing is that the area is not too rugged, allowing for easy movement from one side of the park to another. Second, the park has several picnic areas and good shade, making it a perfect getaway for a family, where you can be on a picnic when relaxing, then out panning for gold the next minute.

The park staff is also often at hand to help everyone select locations and use the available tools when hunting for rocks and treasure.

Herkimer Diamond Mines

Located in New York, the Herkimer Diamond mines are home to some beautiful quartz crystals that have been forming for several hundred million years. These quartz rocks have a diamond-like shape, thus the nickname the 'Herkimer Diamonds.'

These quartz crystals are so beautiful that the mine offers an option to turn your find into jewelry on-site at the Mining & Lapidary Station (Artisan Center).

Speaking of the site, besides rock hounding, the mine also has camping facilities, cabins, fishing, water sports, and many other family-friendly activities.

The Herkimer Diamonds Mine is open all year long but closes in the event of extreme and unsafe weather. Furthermore, the Mining & Lapidary Station is not open on Monday and Tuesday.

Spectrum Sunstone Mine

Located near Plush in Lake County, Oregon, the Spectrum Sunstone Mine is open to the public and offers the Oregon sunstone gem as its main gemstone. The Oregon sunstone is a colorful piece of stone declared a gemstone by the Oregon federal government in the 70s. You can only find this particular gemstone in Oregon. The gemstone's colors include blue, green, and red, with some coming in color-changing forms. Aside from the sunstones, you can also get trilobite and Himalayan tourmaline gems.

The mine has several facilities for those who want to camp there, including cabins with hot showers and flushing toilets. It also has good access roads fit even for small vehicles. The entry fee is $75 for adults, but children aged 12 and below have entry-free access.

Fossil Butte National Monument

The Fossil Butte is in Kemmerer, Wyoming. The location provides a well-rounded rock hounding experience, including a collection of a wide selection of fossils you may never come across anywhere else.

However, the Fossil Butte is very rugged; thus, you need sturdy hiking shoes. There is also minimal shade, and thus, carrying sun protection is recommended.

Fortunately, rock hounding here is very rewarding, and you will be in for a treat, with fossils of plants, insects, reptiles, birds, fish, and mammals up for view.

The park is open throughout the year, but the hours will vary depending on the season.

Flint Ridge

Found in Ohio, Flint Ridge is a top spot for rock hounding in Ohio due to the diversity of the things you can collect from here. You can get a wide range of quartz, amethyst gemstones, carnelian, and chert rock, all of which come in various colors.

The Flint Ridge has ancient quarries you can visit. You can also view the various weapons and tools made by the community that once lived in the location using the available rocks. The park also has beautiful trails through which you can walk and learn about the history of the place.

Trilobite Quarry

Located near Delta, Western Utah, the trilobite quarry is one of the best places to hunt for trilobite fossils. Trilobites are extinct marine arthropods that lived on earth over 500 million years ago in seas that covered the regions around the western U.S. Besides trilobites, you can also collect the fossils of several other sea creatures that have long gone extinct.

The quarry is open to the public from April to mid-October. The cost of collecting the fossils here is $50 for adults and $35 for children aged 8-14, while those who are 7 and under can collect for free. The prices increase if you want to collect higher-quality specimens, but you can get great discounts if you travel in a large group.

The quarry has camping locations within and hotels nearby. During the adventure, you will need sun and insect protection and personal hydration.

Chapter 5: Best Rocks to Collect During Rock Hounding

Rocks are among the most common finds you will come across during your rockhounding expedition; thus, collecting some cool rocks is the best way to make rockhounding a worthwhile hobby.

Rocks do not always have a defining feature that makes them worth collecting, but you can collect rocks based on how rare and cool they look.

Below are some rocks to look out for in your rockhounding adventure:

Oolitic Limestone

Oolitic limestone is a sedimentary rock with a rather distinctive and unique appearance. The rock is made from small ooids made up of calcium carbonate. During their formation, the small ooids (spherical grains of sand particles and other minerals) are washed around on the sea floor, leading to the lime mud in the sea fusing them. The gaps between the ooids are the feature that gives the oolitic limestone its hole-like structures. You will mostly find this limestone in warm, shallow seas or lakes containing a lot of calcium and other minerals present in the seawater. This means the oolitic limestone is quite rare, making it a valuable find.

Locations such as Great Lake in Utah, the Bahamas, and Red Mountain in Alabama are some of the best places to collect oolitic limestone.

Obsidian

The obsidian forms when lava flowing from a volcano rapidly cools. The obsidian is very smooth with a glassy surface. The beauty is what makes the obsidian a must-have rock collectors' piece. While it will often be black, you can find some in a pretty shade of red. **NOTE:** Obsidian is technically not a true rock since it does not have mineral crystals. Instead, we normally refer to it as volcanic glass.

Thunderegg

A thunderegg is a small rock (usually the size of a baseball) formed in the flows of rhyolite lava, a viscous type of lava. Gas bubbles will often form when that lava flows down the side of the mountain. The gas then gets trapped in the rock as the rock cools. Over a long period (thousands of years), small rock fractures allow water laced with different minerals to seep into the rock and fill the gas bubbles within it. Eventually, the rock's center fills with the minerals from the water.

The minerals that fill the void can be made up of quartz, opal, or jasper, depending on the minerals in the water that seeps into the rock. The thunderegg is pretty hard to crack open. You will need a lapidary saw to access its inside.

Geodes

Geodes are rocks considered a must-have for any rock-collecting enthusiast.

A geode is a rock that can either be spherical or subspherical and has a hollow inside lined with minerals. Geodes can form in several ways.

The most widely hounded geodes are volcanic geodes that form when basaltic lava flows develop voids within them, often from gas trapped within the lava. These voids become deposited with minerals like agate, quartz, or opal. However, unlike the thunder egg, the mineral deposits do not fill up the voids. Instead, they line the sides of the void.

Another way geodes form is through sedimentation. Shells, tree branches, roots, or other organic matter might decay in an area made up of sedimentary rocks such as limestone or dolomites. As the organic matter decays on the rock, they leave a void that allows for the depositing of minerals. These cavities can then be lined with minerals, creating a geode. Because the groundwater will often contain different minerals specific to a given location, geodes often have varying interior mineral compositions; this means you will often have different colored insides for different geodes found in different areas. You can find many geodes in regions like California, New Mexico, Utah, Nevada, Iowa, Missouri, Indiana, Kentucky, Illinois, and Ohio.

The geode does not have a distinctive appearance on the outside. Thus, you will often need to understand several things before collecting the rock. These are:

- **The locality –** Understanding the area is crucial. Locations known for producing geodes should be your hunting spots because geodes need specific geological settings to form.
- **Rough spherical shape –** Geodes will often be or almost always well rounded, and those that do not have this shape have some spheroid shape, with some being egg-shaped. They will rarely have sharp angles or pointy edges.
- **Hollow inside –** You can know the insides of a geode by shaking it and listening. Jiggle the rock around your ear and listen. If you hear some jostling inside, then that could be a geode. The movement is the minerals moving around inside the hollow space in the rock. You can tap the rock when you don't hear any movement within; you should hear a hollow sound coming from the rock. You can use another solid rock as your control to tell the differences in the sounds.

Geodes are quite easy to crack open, and you can open one by smashing it with your hammer or another rock against it to confirm your find.

Granite

Granite is a common igneous rock type with a rough texture due to the large grains that make up the rock. The rock forms when magma below the earth's surface slowly crystallizes.

The granite can be white, gray, or pink-colored depending on the minerals that go into making it. It is often made up of quartz and feldspar, though they also have small traces of other minerals such as mica.

Granite is a very common rock and should be quite easy to collect, especially if you visit locations such as Stone Mountain, Georgia; Yosemite Valley, California; Pikes Peak, Colorado; Mount Rushmore, South Dakota; White Mountains in New Hampshire.

Pumice

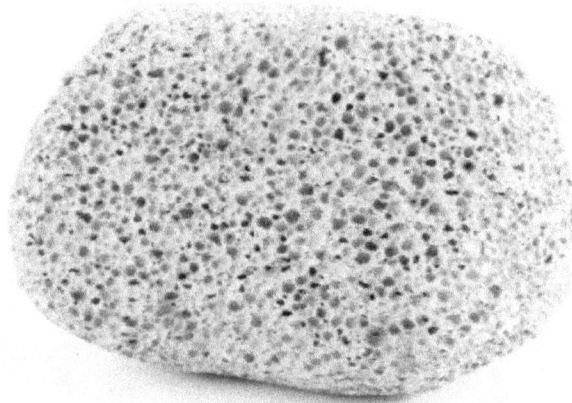

Pumice is a rock with a distinctive characteristic: it is so light that it can float on water. The rock forms from molten magma that was cooled so rapidly that there was no time for crystallization. The numerous pores on the pumice are gas bubbles trapped in the magma during its cooling that remain there due to the fast cooling that left no time for the structure to arrange itself into a solid crystalline structure.

Pumice is common and should be among the first ones you collect during your expedition.

Conglomerate

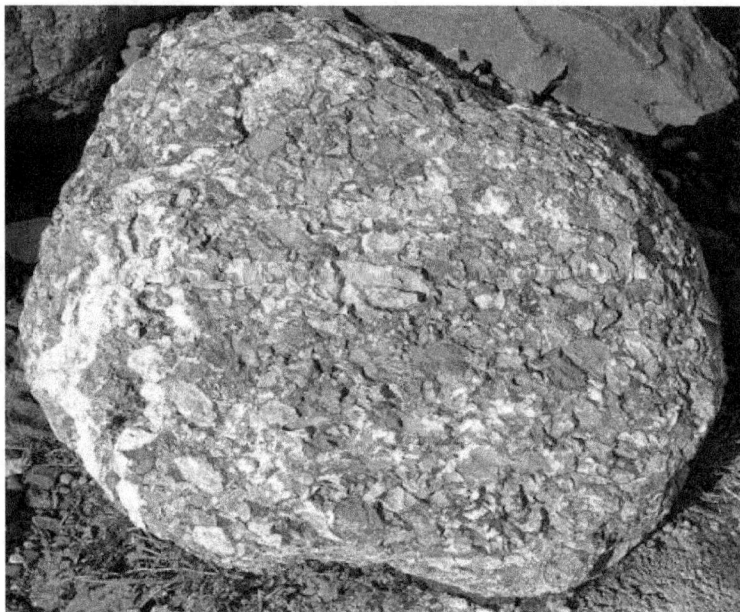

A conglomerate is a unique and interesting rock because it is not a singular rock but several rocks stuck together to make a bigger rock. It is considered a clastic sedimentary rock made up of fragments (clasts) of already pre-existing rock and minerals.

The minerals can be from quartz feldspar or fragments of igneous or metamorphic rocks. It can also contain clasts of limestone, granite, basalt, and sandstone. A mixture of sand, mud and chemical cement such as calcite binds these fragments together. Conglomerates will often form in running water or moving waves. Thus, you are likelier to find them in areas around water bodies or locations that previously featured large water bodies, either a dried lake bed or a dried river bed. Furthermore, beaches or river banks could be great places to pick up a conglomerate rock.

Gastroliths

Some animals use stones to help with digesting their feed. These stones are what we call gastroliths. Animals that do not have suitable teeth for grinding food will swallow these tiny rocks to help grind food in their gastrointestinal tract.

The best gastroliths you can find are those that were from the dinosaurs. So, sites known for dinosaur fossils recovery can be great places to collect these rocks.

Note that we define gastroliths based on animal digestion usage, not their formation. Thus, they can be any kind of rock. However, they will often be very polished and smooth due to the nature of grinding down anything the animal eats.

Tektites

This is another rock that is not technically a rock.

A tektite is a form of naturally occurring glass formed from very high velocity and extreme heat. However, they form from non-volcanic activity.

Tektites are parts of a meteorite. When a meteorite comes into contact with the earth, the high impact and extreme heat will usually melt sediments and rocks. The impact also ejects these sediments and rocks into the air. As the melted rocks fall, they quickly cool down and form the tektite.

Note that the high amount of silica present that gives it its glassy look is the key thing that separates tektites from any other meteorite. So, a tektite isn't just any meteorite part but one rich in silica.

Because they form due to activity from outer space, tektites are probably among the rarest of the rocks mentioned here.

Chapter 6: Best Gemstones to Collect During Rock Hounding

Gemstones are highly-prized and valued minerals due to their beauty, rarity, and durability. Gemstones will often have crystals that make them have much more luster than an ordinary rock would.

Due to their rarity, gemstones will take a lot more effort, time, and resources to find, but they are often worth the effort.

Below are some of the most valuable gemstones you can endeavor to have in your collection.

Opal

Opal is a beautiful gemstone composed of silica spheres arranged in a regular pattern, with water trapped between the spheres. When light diffracts on these spheres, it breaks up into several colors. This property gives the opal its iridescence —or in the case of an opal, its opalescence.

Opals occur in several primary colors; white, black, and red. White opals are the most common and will be the easiest to find. These opals have pale colors. Black opals are rare and often shine brighter with brilliant red, green, blue, and purple colors. Red opals are the rarest and need higher amounts of silica to form.

Emerald

Emerald is a gemstone with a bluish-green color (emerald), which it gets from the chromium or vanadium minerals present. The gemstone has global recognition as being precious due to its rarity. In fact, it is rarer than diamonds.
Most gem locations in the U.S will often produce emeralds in very small quantities, which is why they are not commercially viable, making it possible for rock hounds to collect them. Had emerald been of commercial viability here, there is a high probability you would not be able to collect it for keeps.
North Carolina is the only place in the U.S with any significant emerald deposits.

Tourmaline

Tourmaline is the name of a gemstone that comes in various colors and is a stunning mineral piece. The tourmaline has elongated crystals that often have a hexagonal shape and a translucent appearance.

Tourmaline forms through a very complex and varied chemical formation. However, it needs the presence of a compound called boron, often found in cooling molten rock – magma.

The gemstone primarily comes from an igneous rock called a pegmatite and often develops from water solutions within the magma. Water is necessary because it helps mix the minerals and fills the cracks in the magma, creating conditions ideal for tourmaline formation.

So, tourmalines are gemstones you can find in areas of previous volcanic activity. The easiest area to find tourmaline in the U.S is in the Tourmaline Mines in San Diego.

Jasper

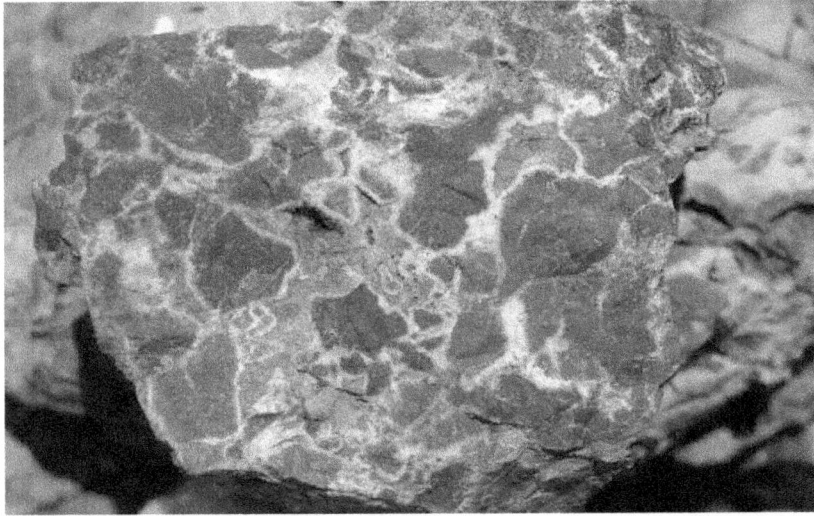

Jasper is a smooth, opaque gemstone that mostly comes in red but has patterned cryptocrystalline quartz, with small quartz crystals that can give it different luster —for example, yellow, brown, white, grey, or even green. Blue is very rare but can also occur. Once again, jasper forms best in volcanic areas, where minerals and sediments rich in silica consolidate volcanic ash deposits.

Here in the U.S, you can find jasper rocks in almost all the states, often around agate rocks, in riverbeds or running streams, on ocean gravel beaches, on hillsides, near mountains, around springs, reservoirs, and creeks.

Fluorite

Fluorite is a mineral that comes in various colors, has beautiful cubic forms, and shines in fluorescent displays under UV light. It often comes in several beautiful colors, such as green, yellow, purple, and blue.

However, fluorite is not a very hard gemstone, which is why it does not find popular usage in the mainstream. Indeed, one way to identify a rock as fluorite is by scratching it with another rock. It should scratch easily, as it has a hardness of just four on the Mohs scale for mineral hardness.

Quartz

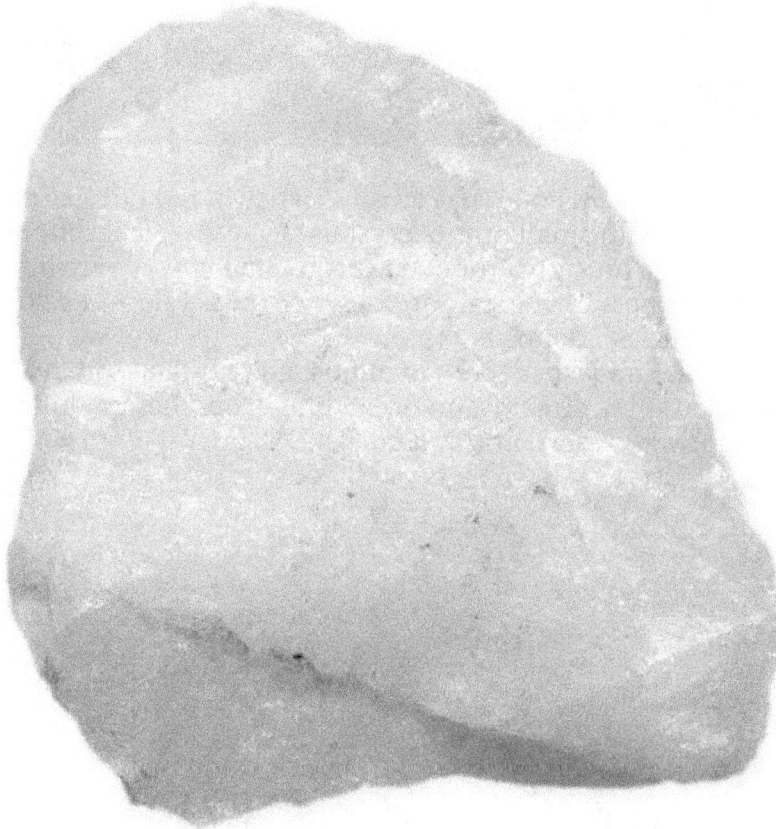

Quartz is an abundant mineral found in almost all rock types found on earth – metamorphic, igneous, and sedimentary rocks. It comes in a wide variety of colors, from white, colorless, green, red, blue, black, pink, purple, brown, orange, yellow, and gray, with some even being multicolored. The purple variety of the quartz is called Amethyst, which is a valuable gemstone all on its own.

Out in the wild, you can identify quartz by a colorless streak, transparent, and very hard crystals with a luster similar to that of glass. It also often comes in six-sided pyramid-like shapes.

Quartz will be easy to find in old mine tailings, soil pockets, or pay-to-dig sites.

Peridot

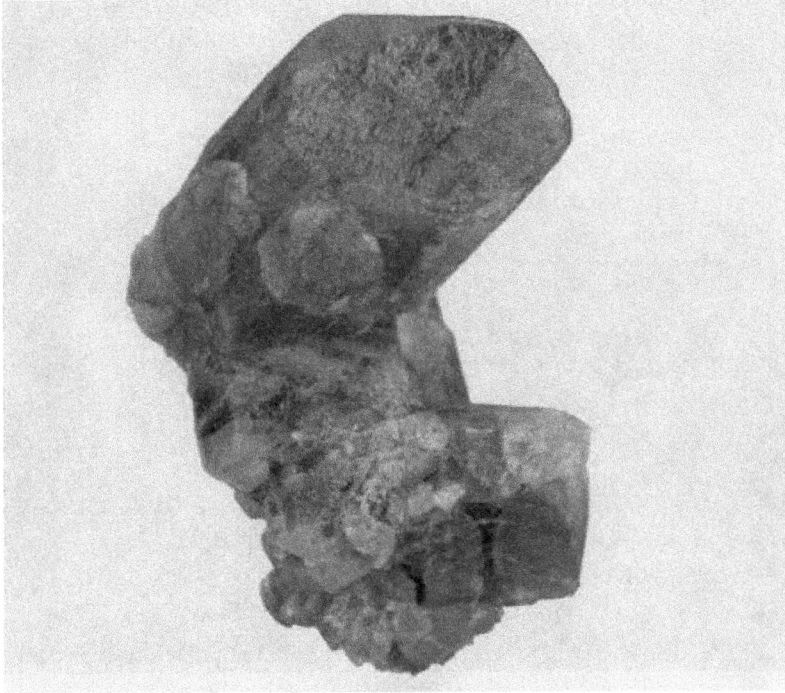

Peridot is a gemstone that usually comes in a deep green or yellowish-green color that forms deep in the earth's mantle, making it a very rare gemstone to come across. Its main components are iron and magnesium, with iron being the cause of its attractive color. Due to its formation very deep in the earth, you are likelier to find the peridot in places that have experienced an eruption from the depths of the earth to the surface.

Garnet

Garnet comes in a reddish shade but can also occur in orange, yellow, green, brown, purple, black, or even pink. Blue garnets are very rare but also occur.

Garnets form when a sedimentary rock rich in aluminum content experiences pressure and heat in a process called metamorphosis. This high heat and pressure break down the rock's chemical bonds, forcing the minerals to recrystallize, forming garnets.

You will most commonly find Garnets in rock outcrops, sandstone locations, rivers, creeks, ocean, and beach gravels are also great places to find them.

Aquamarine

Aquamarine is a gemstone that comes in blue-green but is often more blue than green. The gemstone has a trace amount of ferrous iron, which gives it its distinctive color. Aquamarine forms from pegmatite rocks, which means you can find them with other gemstones made from pegmatite rocks. They form from granite magma that cools slowly. Hot water saturated with metals (mostly ferrous iron) and minerals then move through the veins in the magma as it cools, creating conditions for the formation of aquamarine. Locations with granite and metamorphosis rocks will also have aquamarines; you can also try to find them in open-pit mines. This gemstone isn't too deep in the earth, making it easy to mine with a shove or pick axe once you have identified locations where you can find it.

Topaz

Topaz forms along cavities of igneous rocks such as granite or rhyolite. It forms in the late stages of the cooling magma. As the magma cools and there is a high presence of fluorine, these conditions are perfect for the formation of topaz.

In its natural, pure state, topaz is colorless. However, present impurities can make it a pale yellow to amber. Blue topaz will often undergo artificial treatment because naturally occurring blue topaz is rare.

Areas with a high presence of metamorphic or magmatic rocks are the best hunting spots for topaz. You can also find deposits in river beds and streams, on hillsides and mountains, the most famous of which is Topaz Mountain in Utah. Aside from topaz, you can also find opal and some other precious gemstones on Topaz Mountain.

Chapter 7: Best Invertebrate and Plant Fossils to Collect in Rock Hounding

As we learned earlier, collecting fossils is also a major part of rock hounding as they diversify your collection.

Collecting fossils from vertebrates is prohibited across the United States, but you can still find and collect fossils of other invertebrates and plants without a permit.

Below are some fossils for you to consider in your rockhounding trip.

Invertebrates

Trilobites

Trilobites were arthropods found on earth hundreds of millions of years ago; many people consider them the earliest known group of arthropods.

The trilobites were so common that the fossils can be found preserved in rocks literary anywhere in the world. The most common rocks that perfectly preserve the remains are limestone shales. These shales will often split easily into flat sheets, revealing the fossils.

Ammonites

Ammonites were a group of shelled cephalopods that went extinct 66 million years ago. The shells were tightly wound, which is the state in which we can find the fossilized remains.
You can find these fossilized shells all over the world, \often in sedimentary rocks located in or around marine environments.

Coprolite

This is not the fossil of an insect or a plant, but rather, fossils of feces. This type of fossil is called a trace fossil. A trace fossil is a fossil that records the biological activity of a plant or animal but has no preserves of the plant or animal itself.

You can find these stony masses of fecal matter across many parts of the world, and they often come in various sizes and shapes, depending on the animal. However, if you want dinosaur coprolite, locations near streams, rivers, or waterbeds are the best places to look. Coprolites have the distinctive spiral shape of feces. They also often occur in distinctive dull colors like brown and have a rough, cracked, dusty, or crusty appearance.

Crinoids

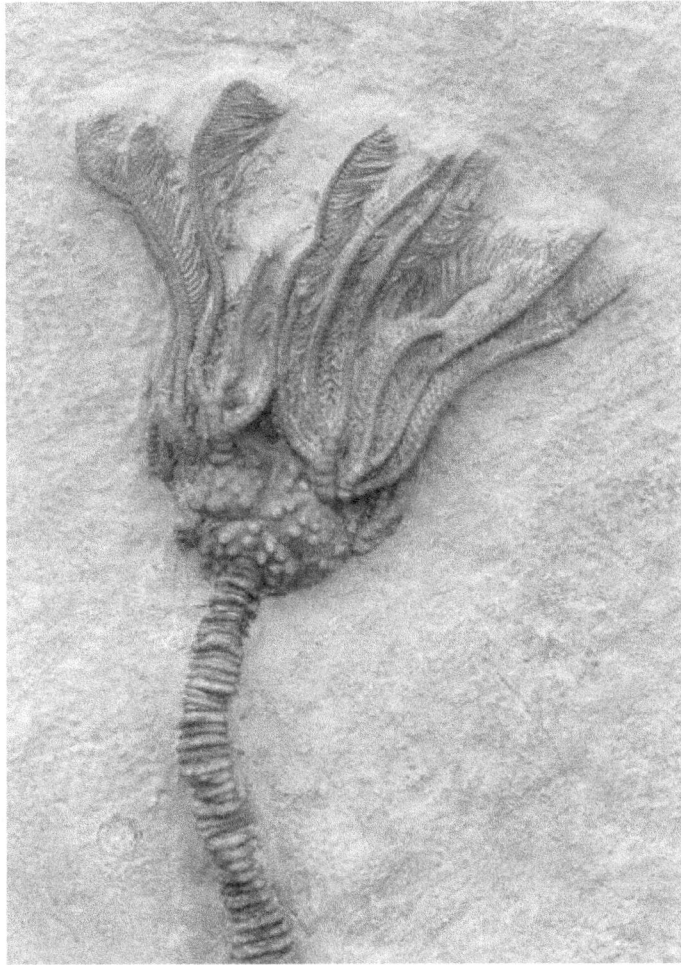

These pre-historic sea creatures are still alive today, though they are now less common than they were during the Paleozoic period, about 280 million years ago.

The earliest crinoids have been preserved in rocks such as limestone shales. They will often be distinguishable due to the arms extending outwards from a small round body called a calyx and a stem extending downwards from the calyx.

These creatures lived in salt water; thus, most exposed marine rocks could contain crinoid fossils.

Megalodon tooth

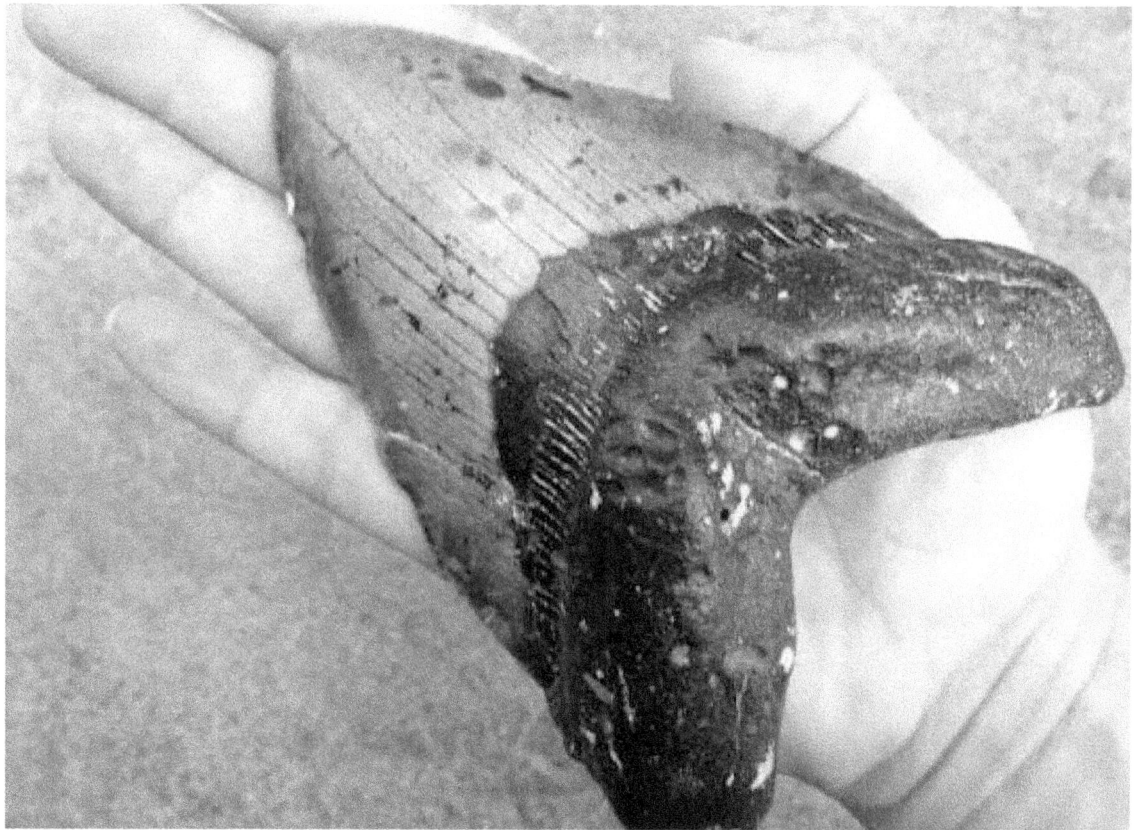

The Megalodon is a pre-historic shark believed to have been the largest sea creature of its time. It is estimated to have lived between 23-3.6 million years ago and could grow to 65 feet (19.812 meters).

Now, while its vertebrates are not for collection, its teeth are. The large and impressive teeth will often be bigger than the palm of an average adult, making it a very impressive find.

You can find Megalodon teeth on beaches in almost all parts of the world; in the U.S, you can mostly find them on the southeastern Atlantic coast extending from Florida, Georgia, and North and South Carolina to Maryland.

Searching along the beaches after a storm increases your chances of stumbling upon this giant tooth from an extinct animal.

Belemnites

Belemnite fossils are the remains of animals belonging to the squid-like cephalopod family. However, unlike the squid, the belemnites had an internal skeleton, which made up the bullet-shaped cone preserved in fossils today. They lived on earth during the period known as the Jurassic and Cretaceous.

These fossils are quite rare, but when found, it is usually in clusters, with their size ranging from a few inches to several feet. Since they lived in shallow water much closer to the shores, you are much more likely to find their fossils on beaches bordering open seawater.

Brachiopods

These are another type of pre-historic creature that is still alive today. Brachiopods are marine invertebrates that live at the bottom of the ocean in a variety of locations in the sea, from the soft sediment deposits, on rocks or reefs, or in rock crevices. Some will even anchor themselves on the ground with a muscle stalk called a pedicle.

The brachiopods fossils found today are of the earliest known brachiopods, which lived around 550 million to 250 million years ago. During that time, the creatures were much more diverse than they are today.

The fossils in focus here are the shells that come in various shapes and sizes. Sometimes, you will find the bottom part, which is convex-shaped, while the top is concave. The outer parts of the shells will often have a covering of concentric wrinkles or radial ribs. Brachiopods also lived in relatively shallow waters (up to 650 feet or 200 meters) and are preserved in shales formed from mud and silt deposits.

Thus, you can find these fossils in preserved marine rocks or beaches. Pennsylvanian rocks in east Kansas are the best place in the U.S to find brachiopods fossils.

Amber fossils

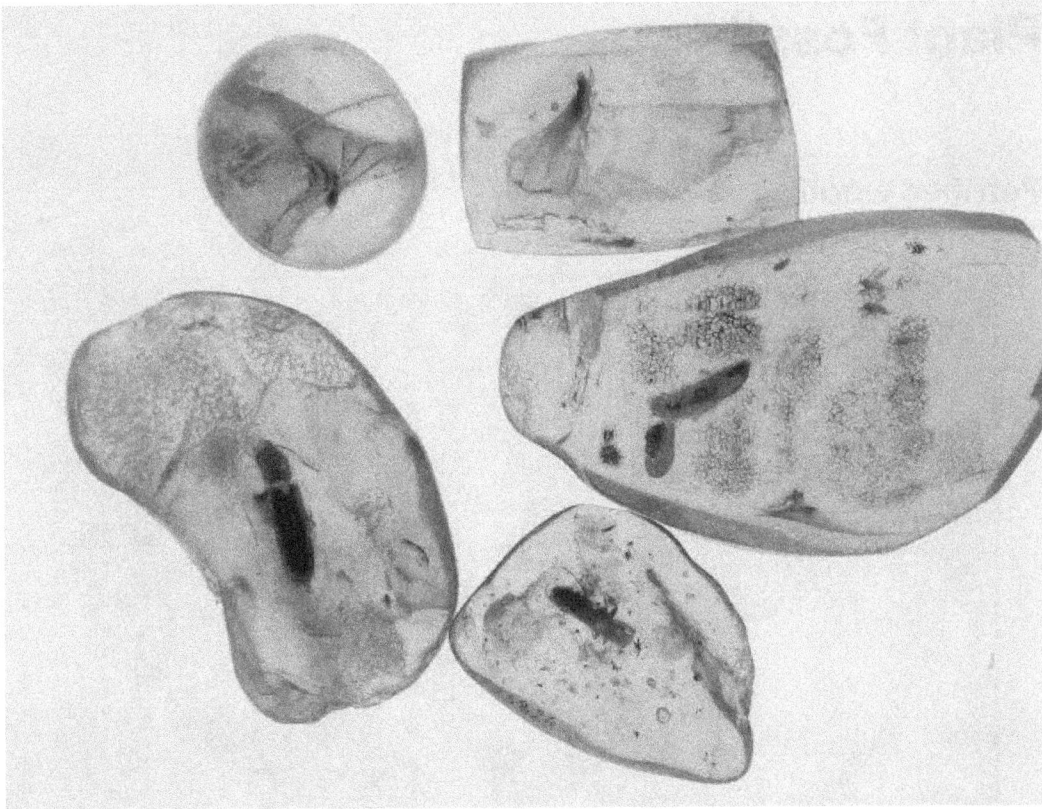

Amber fossils are fossil tree resin preserved over the years due to losing their volatile chemical composition after being buried in the ground.

Amber is created when a tree releases the resin meant to seal its wound. Trees also secrete this thick resin when attacked by insects and fungi, hence the presence of insects and other material in amber.

As the resin trickles, it traps more insects and small parts of plants. For the resin to get preserved as amber, the conditions must be perfect; only then can it become chemically stable and resistant to natural elements. Some of these conditions include needing to be buried in seawater and submerged under layers of sediments. These conditions make it easier for the resin to fossilize.

This is why the Baltic region is the best place to find amber because glaciers in that region destroyed many trees that produced resin. These trees were buried in the region now occupied by the Baltic Sea, and the wet clay and sand sediments in the sea effectively preserved the resin to amber.

Amber will often occur in different shapes and shades of yellow but can also occur in brown, orange, and rarely red.

What makes amber a good fossil sought after by rock hounds is that you will often get species of fossil insects and plants within them. Thus, amber fossils can offer a great view into the past and the insects and plants common during that time.

Plant Fossils

Petrified wood

Petrified wood is among the most common plant fossils to collect. Petrified wood forms when plant material is buried in sediment and protected from oxygen and organisms, thus preventing decay. Then, groundwater rich in dissolved minerals flows through the sediment deposit, removing much of the plant material and replacing it with silica, calcite, or other inorganic matter. So, essentially, petrified wood isn't exactly wood but a mineralized wood fossil —packed with minerals— a process that can take millions of years.
Petrified wood is relatively easy to find, especially in volcanic deposits and sedimentary rocks in several global locations. Petrified Forest National Park in northeastern Arizona is among the best places where you can find petrified wood.

Leaf imprint

A leaf imprint is a trace fossil that usually forms when a plant leaves behind an imprint of its leaf as it rots away with much of the plant not available. Leaf imprints happen when the plant leaf comes into contact with soft rocks or mud, leaving its impression on it.

An example of a leaf imprint is the Pennsylvania fern fossil and the Mississippian fossil plant.

Pennsylvania fern fossils

Pennsylvania fern fossils are common plant fossils to collect on the Pennsylvanian rocks. These fern fossils were present during the Pennsylvanian period (between 323 to 298 million years ago), and you'll often find them preserved in weathered shale and slate found in the local stream.

The ferns will be stored either as impressions in the shale, meaning you don't get the plant part but rather an impression it left on the rock as it rotted away, or a thin film of carbon, meaning there are some original plant cells left on the rock. This type of fossil will often have a detailed view of the plant, including leaf veins.

These ferns provide a glimpse into the period when spore-bearing vegetation dominated the region.

Mississippian fossil plants

Mississippian plant fossils are plant fossils preserved from the Mississippian period, which was between 358 million to 323 million ago.

Plants from this period have been preserved in shales and sandstone. Most of the plants often include seed ferns, ferns that grew to small tree sizes and bore fruit, and scale trees, which were trees with barks that resembled reptilian skin more than anything like a plant.

If you want to find a lot of leaf imprint, consider locations with a lot of sedimentary rocks, which form with mud and sand, making it easier for plants to imprint on them. The low pressure and temperatures through their formation ensure the fossil is not destroyed.

Plant in Amber

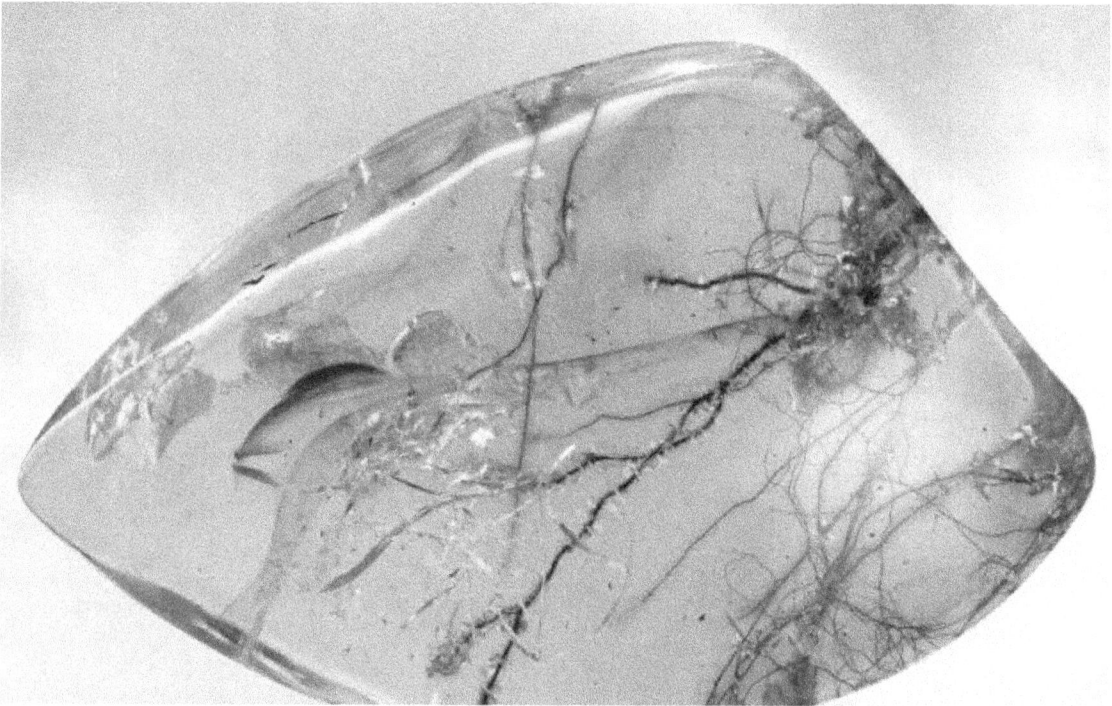

Plants in amber are as described: parts of a plant trapped in amber and preserved. Sometimes, amber might not contain whole plants. Instead, it may have pollen, spores, or flowers.

The best place to find plants in amber is the Central Eastern Atlantic coast of the United States. This type of amber is called the Raritan Amber or New Jersey Amber.

Chapter 8: Tips and Techniques to Use When Rock Hounding

With your tools and other equipment ready, surely there is nothing more to do. It is now time to go out and collect rocks.

However, let us take a moment to look at some tips guaranteed to help make your rockhounding experience much more fulfilling and fruitful.

Use sounds and eyes

When rock hounding, you should be a keen observer. However, as we have seen, some valuable rocks, like the geodes, have no distinctive external features. Thus, you will need to listen to how they sound. Since they are hollow, they should sound hollow, which will differ from how a solid rock would sound.

Always look where you are

Focusing on the rocks around the general vicinity is good, but if you are looking for small rocks or rocks buried under the earth, always keep your eyes focused on the rocks in front of you. Take your time to scan any rock of interest in front and around where you are.

Take the equipment necessary for what you are looking for

You do not need to carry all your equipment when going rockhounding because they could all be heavy enough to bog you down, reducing the area that can cover when out. First, establish what you are looking for. Are you looking for smaller rocks like the thundereggs or geodes? If so, carry smaller tools such as a small rock hammer or a chisel. Are you hounding larger rocks for minerals or fossils? If so, carry your heavier equipment, like the sledgehammer or crowbar.

The same is true for the carrying equipment. If you are hunting for smaller rocks, small boxes and tubes should be the only things you take. If you are hunting larger rocks or think you will take back a huge number of rocks, then a bucket will suffice. Just don't take more than you need to reduce the load you will have to carry.

Identify your trail before leaving

While it is important to identify the region you want to rock hound in, knowing the routes you will take is crucial before setting out. Doing this reduces the chances of getting lost and increases the chances of finding something. Remember, most rock hounds follow certain trails because they lead them to what they want.

That said:
Knowing your trail beforehand will also help you remain oriented even when you decide to branch off into a new trail because you will know your landmarks or identifiers beforehand.

Research the duration of your trip beforehand

Aside from knowing your trail, it is also paramount to know how long it could take to complete your rockhound trip. Rockhound trips much closer to home will often take you a few hours or perhaps a whole day. But when you decide to venture far from home, research the distance you need to travel, how long you can conceivably spend there and come back, and whether there are campsites or hotels around the rock hound location in the event you need a place to spend your night(s).

Use proper hammering techniques on small rocks

When looking to split up small rocks, always ensure you use the proper hammering technique.
First, after wearing your protective tools, ensure the small rock is on a flat surface. Then cup your free hand over the rock with the palm up and the pad of your hands on the rock. This aims to put as much distance between your fingers and palms from the rock as possible.
Then, aim for the center of the rock with the rock hammer and ensure you deal a firm and solid blow through the rock. Use your body for leverage instead of just using your arm strength.
Also, note that different rocks will need different hammering techniques or even different hammers to crack open. So, know the kind of rock you are hitting and its hardness levels before trying to break it up.

Use alarms to keep you on your toes

When rock hounding out in the wilderness, it can be very easy to spend a lot more time than necessary in one location, thus missing out on a lot.
Thus, time management is key.
Set an alarm for at least every hour during your rock hounding. The alarm will be a wake-up call for you to re-evaluate what you have done over the past hour and whether you have been productive.

Make labels

When gathering rocks in the field, make labels that can help you identify what is what when you get back home.

Conclusion

Hobbies such as rock hounding can make life more exciting and worth living because they allow us to dissociate from the hustle of everyday life and enjoy what nature has to offer.

As a hobby, rock hounding needs a lot of planning and resources beforehand. Still, the satisfaction you get from it is worth every effort you put into it. Besides traveling across the U.S gathering valuable rocks, minerals, and fossils, you also get to learn about the history of the places you visit.

By the end of your rockhounding expedition, looking at your rock collection will fill you with pride, seeing all the variety of items you have gathered and the history behind them. Thus, I hope this book acts as a proper guide to you as you pursue this unique hobby. All the best!

www.ingramcontent.com/pod-product-compliance
Lightning Source LLC
Chambersburg PA
CBHW081747200326
41597CB00024B/4426